1893

Duryea brothers make first petrol-engine car in U.S.A.

1896

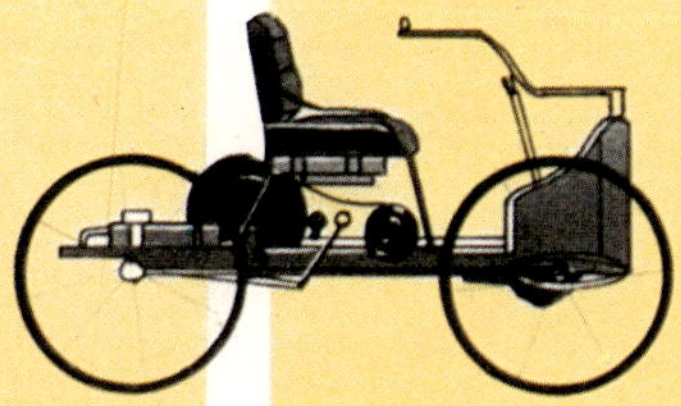

Henry Ford builds his first car

France has most car-makers

1904

Rolls-Royce partnership starts

1908-1927

Over 15 million Model Ts made

1913

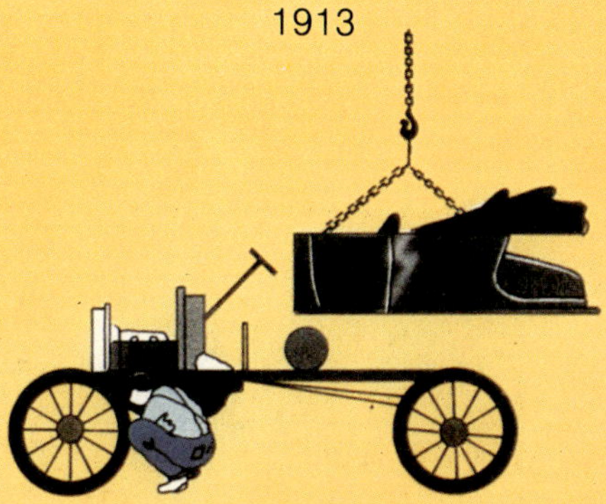

First car assembly line at Ford factory

1959

First Mini made

1980s

Small, economical cars popular

The future

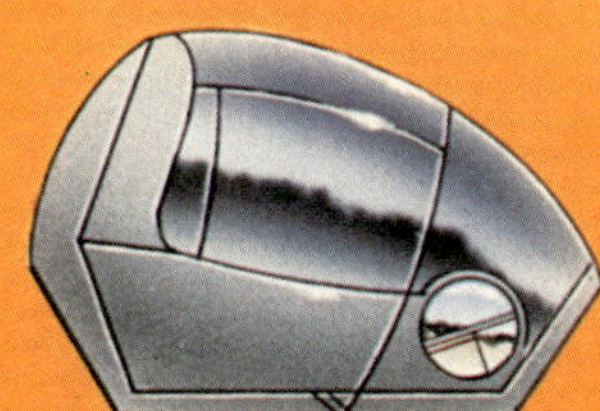

Illustrators

Geraint Derbyshire (John Martin and Artists Ltd) page 8
Gerry Gower (John Martin and Artists Ltd) pages 9, 14–15, 18–19, 24–25
Brian Knight (John Martin and Artists Ltd) pages 16–17
Andy Miles cover background spots, end sheets
Mike Mills pages 22–23
Tom Stimpson cover feature, pages 28–29
Peter Visscher pages 6–7, 10–11, 12–13, 20–21, 26–27

Editor Sue Tarsky
Designers David Bennett
Matthew Lilly

Artists' reference

cover, end sheets, title page, pages 10-11, 12-13, 14-15, 18-19, 22, 24-25, 26 The National Motor Museum, Beaulieu; pages 6-7 BBC Hulton Picture Library; page 15 'The Spirit of Ecstasy' mascot reproduced by permission of Rolls-Royce Motors Ltd.; pages 20-21 Ford Motor Company; page 24 Portman, sole concessionaires (Lamborghini)

First published 1984 by Walker Books Ltd
184–192 Drummond Street
London NW1 3HP



Printed and bound in Spain
by Artes Gráficas Toledo, S.A.
DL-TO-128-84

British Library Cataloguing in Publication Data
Lane, Andrew
World on wheels.— (The Story of the car; 1)
1. Automobiles—Juvenile literature
I. Title II. Series
629.2'222 TL147

ISBN 0-7445-0116-4

# CONTENTS



For Adam and Helen

A.L.

# WORLD ON WHEELS

Written by Andrew Lane

WALKER BOOKS
LONDON

# Starting up

One hundred years ago, a new sound was heard. It was the 'chitty, chitty' noise of a car engine. Today, everyone knows what a car sounds and looks like, but people were amazed by the first car. City streets looked very different before the car. Roads were just stony tracks.

There were horses everywhere, pulling coaches, carriages, carts and wagons.

Some of the first ideas for horseless vehicles were ridiculous. People tried a carriage pulled by flying geese, and even a coach with sails.

1834 L'Eolienne

When steam power was discovered, a steam engine was put on to a wagon in France. It was the first road vehicle to be moved by an engine, but it went out of control and crashed into a wall. In a steam engine, fire heats water to make steam. It is released and its power works the engine.

1769 Cugnot Steam Wagon

1833 Walter Hancock's Autopsy

Steam engines were dangerous. The steam had to be released properly or the engines could blow up, sending steam and metal into the air. Two men used steam buses in England, but they were slow and heavy.

The bicycle was popular in the 1890s, but it didn't have enough power to solve transport problems. It did give ideas to future car-makers. They copied the air-filled tyres and spoked wheels, which were strong but lightweight.

1880 Rover Safety Bicycle

# Out on their own

When Karl Benz first appeared on a spluttering, three-wheeled vehicle, the local people were shocked. They wondered how the vehicle moved without a horse, and laughed when it stopped and Benz had to push it home. The jeering crowd had just seen the first motor car.

The Benz engine used petrol as fuel. Until then, petrol was used to take stains off clothes. Benz found that petrol would blow up when lit. Explosions in the engine gave the power to drive the back wheels.

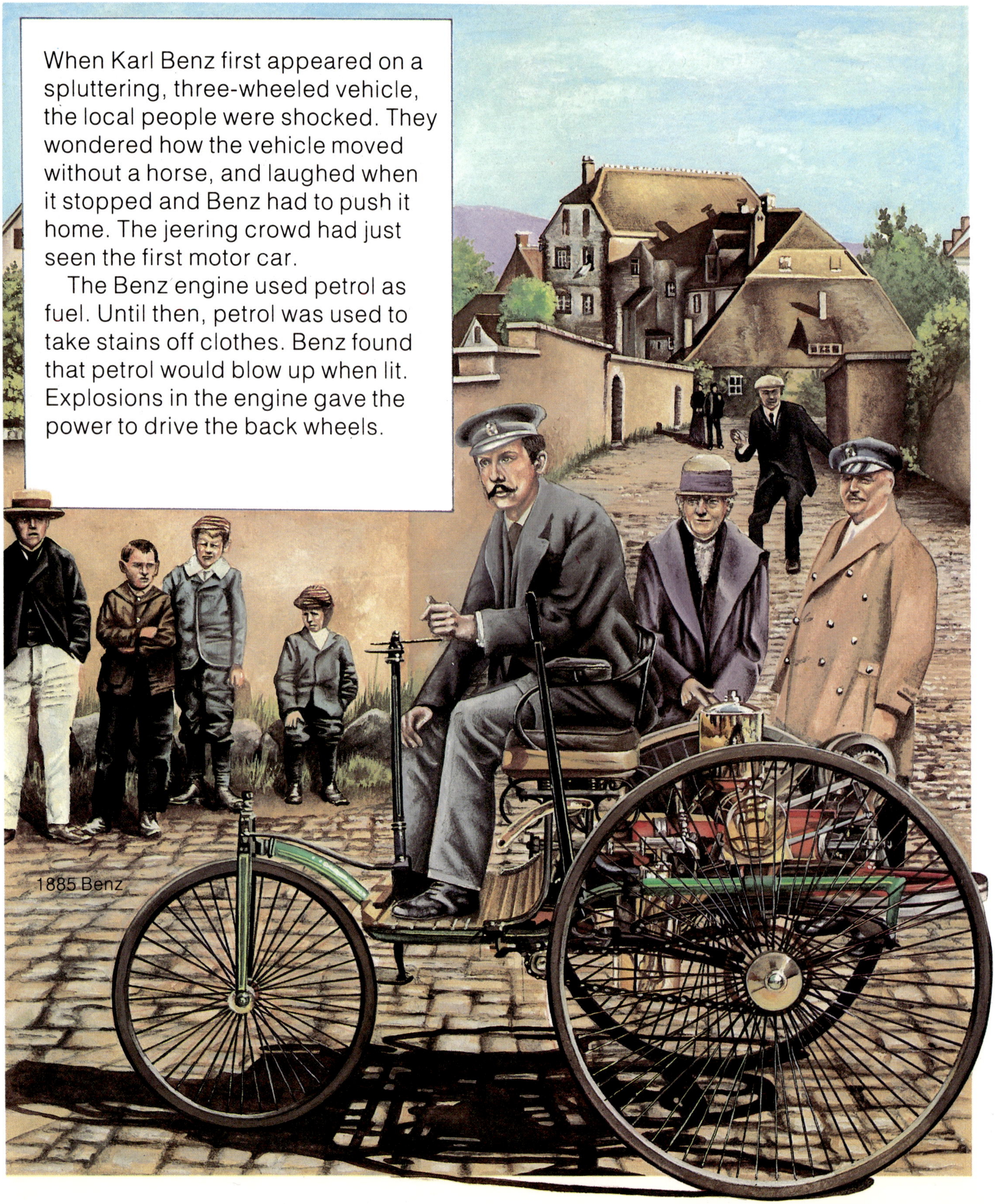

1885 Benz

At the same time, in another part of Germany, Gottlieb Daimler made a better petrol engine. The two men did not know each other. Daimler put his engine in boats, carriages and airships. He put it in a bicycle to make the first motor-bike, but it caught fire when his son rode it.

# Fans and critics

Benz began making and selling cars, and Daimler sold his engines all over Europe. In France, Count de Dion, the Renault brothers, René Panhard and Emile Levassor started making cars. The French people loved cars, held the first races and soon had more car-makers than any country.

1903 Renault

Cars built before 1919 are called Veteran. They were hand-made and expensive. The 1891 Panhard-Levassor set the future style. Its engine was under a front bonnet, a gearbox gave different speeds and a chain from the engine turned the back wheels.

In England, most people hated this noisy, smelly nuisance. Cars went too fast, ran over chickens and scared the horses. Horsemen even whipped any passing cars. Before 1895, a man with a red flag had to walk in front of a car to warn people that it was coming.

1902 Darracq

In the U.S.A., cities were far apart and there were very few roads. It took a long time for cars to become popular. When the Duryea brothers made the first petrol cars there, one was shown as a freak in a circus.

# A day out

1

One day in 1905, a family goes out for a drive. It is an adventure – they never know what will happen. They wear clothes that protect them from cold, dust and rain.

2

1902 Panhard

There are no doors, windscreen or roof to protect motorists.

5

Slowly! The speed limit is 32km per hour and the brakes aren't very good.

6

Oh no! A puncture from a horseshoe nail. Out comes the repair kit.

8

Lunch time! Country picnics are easy to get to in a car.

9

A quick stop at a hardware shop to buy oil for headlamps and a can of petrol.

3

A crank handle is turned quickly to start the engine, a difficult job.

4

The driver jumps into his seat and releases the hand brake. They're off!

7

On the way again. Look! There's an old car that's broken down. Its brakes are so bad that it has a sprag, a metal bar that drops from the back of the car to help keep it from rolling.

10

Which way? A few good sign posts and new road maps help show the route.

11

Time to go home before it gets dark. Oil headlamps don't give much light for night-driving. What an exciting day in the motor car!

# Rolls meets Royce

Henry Royce was a poor English boy whose first job was selling newspapers. He worked hard and finally had his own factory. He began making a few quiet and reliable cars that were better than the other cars made then.

Charles Rolls was wealthy, a racing driver and owner of a car showroom. He went to meet Royce and they decided that Rolls would sell the cars that Royce made. Their 1906 Silver Ghost was soon called the 'Best Car in the World'.

1979 Rolls-Royce Silver Shadow

Rolls-Royce cars are famous for beauty and craftsmanship. They are safe and strong. Since 1911, all the Rolls-Royce cars have had the well-known mascot 'The Spirit of Ecstasy' on the top of their radiators.

1906 Rolls-Royce Silver Ghost

# The car at war

1915 Rolls-Royce Armoured Car

During World War I, some Rolls-Royce cars were turned into military vehicles – some even had 4,000 kilograms of armour added. Before then, people didn't think that cars could be useful in battle.

At first, horse transport was used near the fighting while ambulances, staff cars and lorries were used away from it. But this soon changed and tanks and armoured vehicles with machine guns were used.

1944 Ford Jeep

By the time World War II began, armies from each country had many kinds of military vehicles. The American Jeep was the most outstanding. Each wheel was driven separately, so it could go over obstacles, and the eight gears gave the engine extra power. This meant that the Jeep could go almost anywhere – through soft desert sand, thick mud or overgrown jungles.

# The roaring twenties

1923 Citroën 5CV

Cars of the 1920s are called Vintage. Like earlier cars, they were expensive because each was made to order by hand. But they were better than Veteran cars, with roofs, doors, windscreens, strong tyres and good brakes. Most important, the cars now had electricity, which changed motoring by making it safer and easier.

1928 Hispano-Suiza

There were electric windscreen wipers, headlights, brake lights, indicators and starters. Driving, short hair and dancing to jazz became very fashionable with wealthy young women. There were few cars on the road, not many traffic rules and no driving tests. It was a care-free time for motoring.

# A car for everyone

The American Henry Ford built his first car in a shed and had to knock down a wall to get it out. Ford made more cars and finally had his own company. He wanted to make an inexpensive car that anyone could buy, so he began the first mass-production factory to make the Model T. There was a moving assembly line. Each worker added one part before a car moved on, so the car was made quickly and cheaply. By 1922 about half the cars and vans on streets were Model Ts. Ford had changed the world of motoring.

Molten iron was tipped into moulds for the engines and wheels.

Finished wheels came down the moving assembly line.

Car bodies were fitted to the chassis.

Windscreens and soft hoods were fitted to the cars.

In Ford's factory, men loaded iron and rubber from ships to trains.

The trains carried iron to the steel works.

Engines were fitted to the chassis, or base-frames, of the cars.

Car bodies were spray-painted black – the only colour in which the Model T was made.

A finished car drove away every 10 seconds. In all, over 15 million Model Ts were built and sold. Ford's dream came true – he put the world on wheels.

# On the road

By the 1930s millions of people had their first cars. They were able to move to the new suburbs and drive to work in towns. The first traffic lights were put up.

1935 Ford V8

Car-makers in England, France, Germany and Italy had mass-produced their own inexpensive cars. Garages and petrol pumps were common. People could go where they liked, anytime they chose. They took motoring holidays and drives to the seaside. The car had replaced the horse.

1934 Lancia Augusta

Motorists had to take driving tests. Roads were well-made and highways were built in the U.S.A., Germany and Italy. Drivers travelled long distances easily.

1936 Ford V8 'Woody'

1934 Ford V8

1937 Austin Cambridge 10 Saloon

# Sports cars

Drivers who want cars to look special and go fast often choose sports cars. Early sports cars had open tops. A 'convertible' roof made of canvas was used in bad weather, but was usually folded back. Today, sports cars have streamlined, or smooth, body shapes.

1973 Ferrari Dino 246 GTS; Italian; top speed 225 km per hour; a low, streamlined shape.

1983 Lamborghini Countach LP 500 S; Italian; top speed 300 km per hour; the fastest sports car ever made.

1930 M.G. M type ; English; top speed 115 km per hour; a classic early convertible.

1957 Mercedes-Benz 300 SL Gullwing Coupe; German; top speed 240 km per hour; the doors opened upwards.

1971 Jaguar E-type; English; top speed 240 km per hour; popular and well-made.

# Popular cars

Today, there are millions of cars all over the world. The car is an important part of our lives. People even give cars nicknames, like the Beetle. Over 200 million German Volkswagen Beetles have been sold – more than any other car. It is well-made, reliable and inexpensive.

German Volkswagen Golf GTi

Italian Fiat Panda 45

English Ford Fiesta Popular

Everyone – students, grandparents, families and royalty – likes the Austin Mini. It is fun to drive and has good space inside for its size. The Mini made small cars popular and, today, many car companies make small, economical cars that help save petrol and are inexpensive to run.

Japanese Honda Civic 1300

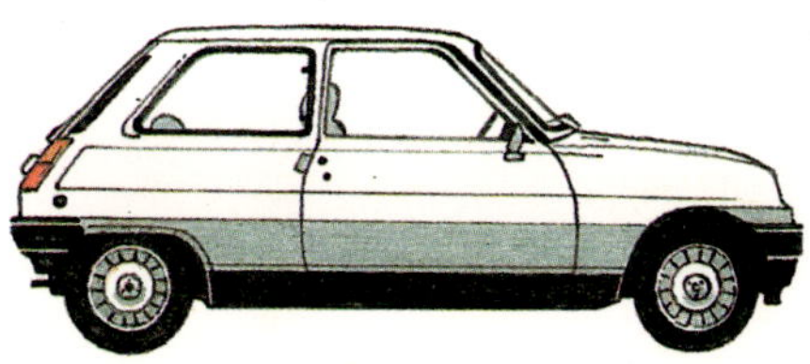

French Renault 5

English Austin Mini Metro 1.0 HLE

# The future

No one knows exactly what cars will be like in the year 2500, but car-makers can tell us some things. Cars will be made of strong plastic, so lightweight that it will be possible for drivers to have a new kind of car, a hover car. While you glide round above the ground, passengers will be able to play video games on their very own computers. Other computers may talk to you, warning of mechanical faults.

Cars could run on lightweight, long-lasting batteries, or even on laser beams. Motoring may be safer than it is today – with built-in radar to control your speed, steering and braking, there may be no accidents. Dashboards could have screens showing charts, flashing digital figures and graphs. One thing is certain – no matter how the car changes, it will still be important to us in the future.

1769

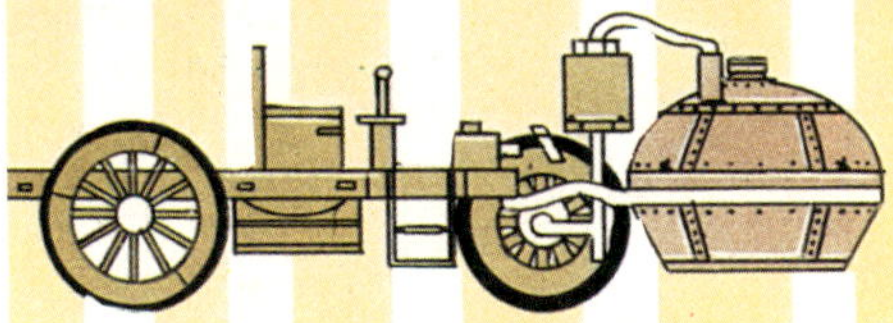

Nicholas Joseph Cugnot builds first mechanical road vehicle

1885-1886

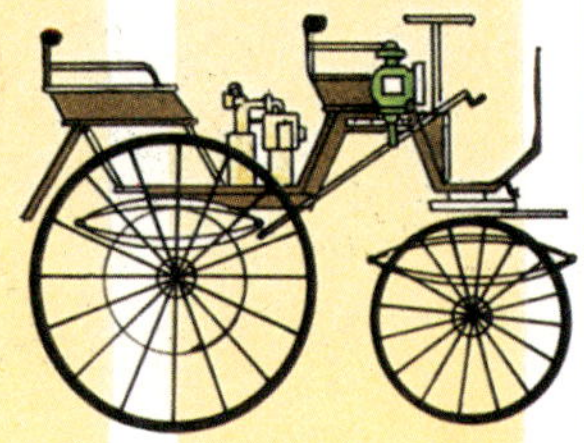

Karl Benz and Gottlieb Daimler make first petrol-engine cars

1891

Panhard-Levassor car sets style for future cars

1914-1918

Cars used in World War I

1920s

More women driving cars

1930s

Inexpensive-car boom

1939-1945

Military vehicles used in World War II

1945

Production starts on Volkswagen Beetle

1950s

Motoring part of everyday life